Jazzy Colors

by
Sonja McGiboney

ISBN: 978-1-7333663-0-4

Jazzy Colors

Jazzy loves her colors,
And she makes it very clear
All the ones that she loves best
For each month of the year.

GORILLA!
JAZZY
Orchard Books
FEBRUARY
MARCH
APRIL
JUNE
OCTOBER
NOVEMBER

January

On cold days in winter,
After snacks and a hug,
Jazzy lies down
On her fluffy, white rug.

February

On Valentine's Day,
There's a cake with a rose.
Jazzy takes a bite,
Getting red on her nose.

XO
XO
OVE!

March

On a warm day in March,
She is sure to be seen
Wearing fancy clothes
Like a queen, in green.

April

As bunnies bounce around
And the birds chirp cheers,
Jazzy sits calmly
Wearing pink rabbit ears.

May

On the fifth day of May,
On a blue, striped mat,
Jazzy falls asleep
With her big, gold hat.

June

On the bank of the river
Where the grass grows tall,
Jazzy likes to play
With her blue beach ball.

July

In the middle of the summer
When the heat leaves you
puffing,
Jazzy stays inside pulling
Purple pillow's stuffing.

August

Lurking in the grass
(Don't worry, it's a fake),
Jazzy keeps on staring
At the skinny, black snake.

September

Gazing out at Mason Street,
Through the window's glass,
Jazzy brightens up to see
The yellow school bus pass.

October

When you see the bats and cats
And witches start to rove,
Jazzy joins her little friends
Around an old gray stove.

November

When the feast is over
And her hunger satisfied,
Jazzy waits to eat dessert:
An orange pumpkin pie.

December

Lying with her reindeer
By the festive white spruce,
Jazzy kisses her new friend,
A fluffy, brown moose.

Sonja McGiboney grew up in Stowe, Pennsylvania. After obtaining her undergraduate degree in music from West Virginia University, she married Dale and accompanied him on his 25-year military career. She has two wonderful children, Rachel and Ryan, and now lives with Dale and Jazzy in Smithfield, Virginia. She loves taking photos and writing Jazzy's Books.

Check out all the Jazzy's books!

1. ABC Jazzy
2. Counting Down Jazzy
3. Growing Up Jazzy
4. Hide and Seek - Jazzy's Alphabet Adventure
5. Jazzy and Friends
6. Jazzy Colors
7. Jazzy Time
8. Jazzy Shapes
9. Jazzy Explores the Library
10. Jazzy Explores Smithfield, VA
11. Jazzy Explores Murfreesboro, NC
12. Jazzy's Halloween - A Night in Ghouling Brook
13. Jazzy's Twelve Days of Christmas
14. Little Red Jazzyhood
15. Princess Jazzy - How to Prove You're a Princess

Made in the USA
Coppell, TX
05 November 2026

11302983R00019